All About Turtles

By Fatima Toure

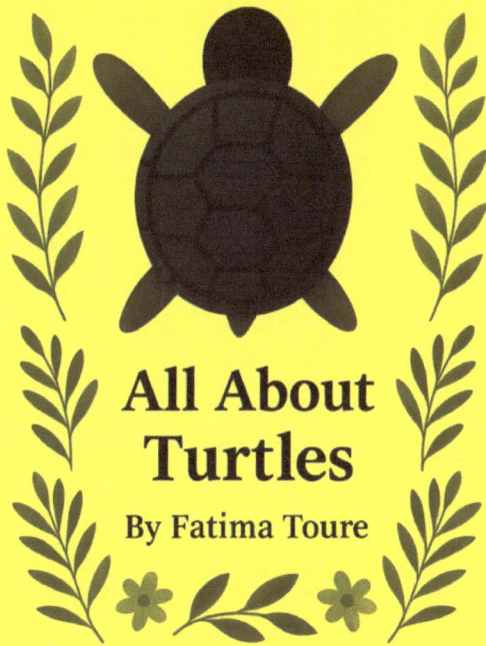

BabouConnect Publishing

All About Turtles

Copyright © 2025 Fatima Toure

Illustration and design by Fatima Toure

ISBN:979-8-9939072-8-4

Published by:

BabouConnect Publishing

sabilfalah@gmail.com

Dedication & Acknowledgment

A Fatima Book 🐢 for every child who loves to wonder.

This book began as a classroom project.
With the help of teachers, family, and a lot of curiosity,
it turned into something bigger — a story to share with the world.

**Special thanks to the school where young minds grow
and dreams begin.**

Foreword

🐢🐢🐢

A Light That Guides

One evening, Fatima and I were talking about baby sea turtles.
She told me how the moonlight can confuse them — how, instead of crawling to the sea, they might follow the wrong light and go the wrong way.

I smiled and laughed, just imagining turtles going sideways.
But Fatima turned serious and said,

"No. It's not funny... because they'll get picked up by seagulls."

That moment stopped me.

In those words, she reminded me that even nature's smallest creatures face danger when they lose their way —
and that stories are not always for laughs. Some stories are about **care, compassion, and survival.**

This book may be about turtles,
but it's really about a child who sees the world through **gentle eyes and a strong heart**.

Her name is **Fatima Toure** 🐢
And I am proud to be her father.

     Mahamadou S. Toure

Contents

Something amazing is coming... get ready to dive in

Chapter 1
Turtle Shell Development.

🎧 Scan this code to hear
Fatima read the chapter!

This chapter is all about turtle shell!

A sea turtle's shell is a hard dome that helps protect it from predators.
The top of the shell is called the carapace.
The bottom of the shell is called the plastron.
Turtle shells are important—they grow with the turtle!

I hope you have learned all about turtle shells

Chapter 2
Where Turtles Live

🎧 Scan this code to hear
Fatima read the chapter!

This chapter is all about where turtles live

Sea turtles live in warm oceans all around the world.

They like shallow waters where they can find food,

and sometimes they come to the beach to lay eggs.

Some turtles live in lakes, rivers, and even swamps! Thet's always stay close to water —it helps them swim, eat, and stay safe.

Isn't it amazig how turtles always find their way home?

Chapter 3
What Sea Turtles Eat.

🎧 Scan this code to hear
Fatima read the chapter!

This chapter is all about what turtles eat!

Sea turtles eat jellyfish, kelp, seagrass, and even small fish. They don't have teeth, but their strong jaws help them bite anıd chew. Some turtles hunt alone, gliding through the water quietly.

Others hunt in groups, swimming together to catch their food.

Different turtles like different meals—Green Sea turtles love plants, while loggerheads ρrefer crabs and clams!

Turtles are amazing eaters.

I hope you've learned all about what sea turtles like to eat

I hope you've learned all about what sea turtles like

Chapter 4
Sea Turtle Babies

🎧 Scan this code to hear
Fatima read the chapter!

This chapter is all about turtle babies!

Did you know that sea turtles lay about 100 eggs in a single nest?

The mother turtle crawls onto the beach at night, digs a hole, and gently hides her eggs in the sand.

After several weeks, the baby turtles hatch — but only at night!

They follow the light of the moon and stars, crawling quickly toward the ocean.

It is a race for survival.

I hope you've learned all about sea turtle babies 🐢

Chapter 5
Fun Turtle Facts.

🎧 Scan this code to hear
Fatima read the chapter!

This chapter is all about turtle facts!

Turtles live a long life. Some turtles can live from 30 to over 100 years!
Turtles sleep underwater. They can hold their breath for a long time while resting.
They move slowly on land...
 But once they reach water — zoom! They swim fast!
Turtle babies hatch only at night!

I hope you have learned all about turtle facts.

Reflection Page

🐢🐢🐢

From the Classroom to the World

This book started as something small.
A classroom project. A love for turtles.
A little girl picking pebbles in the backyard,
Asking questions, making discoveries.

But when love and curiosity come together,
They grow into something *big*.

This book is not just about turtles.
It is about **seeing beauty**, **asking why**, and **sharing what you learn**.
It is about remembering that every big journey starts with a small question:

"What about me?" 🐢

This book is Fatima's answer.
And today, the whole world is listening.

Scan the QR code to listen to the story

Master Conclusion

🐢🐢🐢

From Pebbles to Pages: The Journey of Wonder

This book began with a child's quiet love for turtles—
Shaped by the backyard, by pebbles, by aquariums,
And by a voice that asked, *"What about me?"*

And from that question rose a story,
A drawing that became a screen saver,
A classroom project that became a real book.

This is more than a book about turtles.
It is a book about seeing the world with gentle eyes.
A book about curiosity, beauty, and belief
About a father who saw the light in his daughter's hands
And said, *"Let's take this to the next level."*

To every child who wonders,
To every parent who listens,
To every pebble that becomes a story,
This book is for you.

And to **Fatima Toure**—🐢
Your voice matters.
Your stories matter.
And now… the world gets to read them.

The journey has just begun 🐢

—Mahamadou S. Toure

Father's Reflection

Scan the QR to listen

About the Book

All About Turtles is a joyful journey through the life of turtles – written and illustrated by young author **Fatima Toure.**

Inside, you'll learn:

- How turtle shells grow and protect them
- Where turtles live and sleep
- What turtles eat (yes – jellyfish!)
- And fun turtle facts you'll love to remember

Each chapter includes a QR code so you can hear Fatima read it aloud in her own voice!

Scan & Listen!

Each chapter includes a QR code so you can hear Fatima

www.ingramcontent.com/pod-product-compliance
Lightning Source LLC
Chambersburg PA
CBHW040812300326

41914CB00065B/1516